Marlisa Bernardi de Almeida
Luciana Peron
Ricardo Desidério

School assessment from the point of view of maths teachers and students

Marlisa Bernardi de Almeida
Luciana Peron
Ricardo Desidério

School assessment from the point of view of maths teachers and students

Conceptions of assessment held by maths teachers and students in the final grades of primary school in the public school system/PR

ScienciaScripts

Imprint

Any brand names and product names mentioned in this book are subject to trademark, brand or patent protection and are trademarks or registered trademarks of their respective holders. The use of brand names, product names, common names, trade names, product descriptions etc. even without a particular marking in this work is in no way to be construed to mean that such names may be regarded as unrestricted in respect of trademark and brand protection legislation and could thus be used by anyone.

Cover image: www.ingimage.com

This book is a translation from the original published under ISBN 978-613-9-66058-2.

Publisher:
Sciencia Scripts
is a trademark of
Dodo Books Indian Ocean Ltd. and OmniScriptum S.R.L publishing group

120 High Road, East Finchley, London, N2 9ED, United Kingdom
Str. Armeneasca 28/1, office 1, Chisinau MD-2012, Republic of Moldova, Europe
Printed at: see last page
ISBN: 978-620-7-98797-9

SUMMARY

Marlisa Bemardi de Almeida[1]

Luciana Del Castanhel Peron[2]

Ricardo Desidério[3]

Ellen Bemardi[4]

SUMMARY

The aim of this study was to investigate the conceptions of learning assessment held by maths teachers and students in the final grades of primary school in the public school system in the state of Paraná, as well as other important issues relating to present assessment practices

in the teaching and learning process. The choice of a qualitative study made it possible to analyse the answers to the questionnaires applied to teachers and students at this level of education, seeking to categorise them and associate them with a theoretical framework appropriate to each analysis made. The results showed that the conceptions were often associated with the verification of learning, primarily serving the teacher. These results, far from being conclusive, nevertheless broaden our understanding of

[1] Graduated in Mathematics and Pedagogy, specialising in Psychopedagogy from UNICENTRO. Master's student on the Master's Programme in Education for Science and Mathematics Teaching at the State University of Maringá (UEM). marlisabemardi@yahoo.com.br
[2] Graduated in Maths from UNIOESTE, specialising in Psychopedagogy from UNIVALE. Master's student on the Master's Programme in Education for Science and Mathematics Teaching. lucianaperon@hotmail.com
[3] Degree in Science, specialising in Mathematics Education from UENP/FAFICOP. Master's student on the Master's Programme in Education for Science and Mathematics Teaching, rickdesiderio@hotmail.com
[4] Degree in Chemistry from UNICS. Maths graduate from PRÓ-MINAS, Master's in applied chemistry from UNICENTRO. small_don@hotmail.com

the views of teachers and students by providing a set of very relevant information, both from the point of view of theoretical reflection and from the point of view of educational practice.

Key words: *Learning assessment. Teachers and students. Teaching and learning process.*

CHAPTER 1

INTRODUCTION

Through the history of assessment in everyday school life as elucidated by Luckesi (2002), we realise that the practice of learning assessment that has been developed in many educational institutions leads us to a position of little progress. It has generally not been used as an element that helps in the teaching and learning process, getting lost in measuring and quantifying knowledge, failing to identify and stimulate individual and collective potential.

According to Santos (2008), issues relating to the teaching and learning of maths are the order of the day. Various international documents have a strong influence on our country. From an absolutist conception of maths, which sees it as an objective, fixed, certain, neutral body of knowledge (ERNEST, 1991 apud SANTOS, 2008), a fallibilist perspective of maths stands out, more interrelated with problem-solving since it is seen as a developing human creation and invention. From a list of rules and properties, maths is understood as a science of 2 patterns. Teaching is no longer seen as mainly the rigorous transmission of information, but primarily the construction of situations in which the student can get involved in order to develop their mathematical competence. Learning is not the absorption of fragmented information, resulting from repetitive practice, but occurs through rich and meaningful mathematical experiences. "Knowing maths is doing maths".

Santos (2008) goes on to say that it is one thing to prescribe guidelines, but quite another to apply them in practice. In order to respect the principle of assessment as an integral part of the teaching and learning process, and at the same time develop an assessment that focuses on what is now understood as mathematical competence, a multiplicity of assessment processes can be applied.

In their studies, Moraes and Moura (2009) state that the process of humanisation depends on the objective conditions for man to appropriate the goods produced historically and socially by humanity. In this sense, researchers in cultural-historical psychology, including Vygotsky (1989, 2000, 2004), Leontiev ([197-], 1983, 2001), Luria (1996) and Davídov (1982, 1988), defend education as a universal form of human development. From this perspective, the school is considered the space par excellence for the development of scientific concepts, the institution capable of mediating between everyday and scientific concepts. Saviani (1991, apud to MORAES E MOURA, 2009) states that: "In short, through the mediation of the school, there is a passage from spontaneous knowledge to systematised knowledge, from popular culture to erudite culture". In order for school work to be able to mediate between the knowledge that students have and the theoretical knowledge that has been historically elaborated, it is necessary to organise teaching appropriately and, consequently, to assess learning in such a way as to be able to express whether the school is fulfilling its main function - to enable students to appropriate theoretical knowledge. According to Moraes and Moura (2009), Leontiev ([197-]), based on Vygotskian assumptions, centred his studies on understanding human activity. As the main researcher of the Psychological Theory of Activity, he sought to clarify that consciousness is formed with and in the practical activity of men, as a product of the development of objective relations. The structuring elements of activity are: need, motive, action and operation. The motive is governed by a need, which mobilises actions, which are subordinated to objectives and depend on the conditions for their realisation through operations, which are nothing more than ways of carrying out the action. Activity only exists through actions or groups of actions; the same action can be part of different activities. And the same motive can be realised in different objectives and generate different actions. Activities are processes characterised by constant transformations, always in movement; an activity can become an action and vice versa. Real motives respond to human needs, which are historical. Motives can be material or mental. In a cognitive activity, motives and actions are mental. For example, a schoolboy reading a book is a cognitive activity, but it is carried out through external actions that can take the form of mental processes.

In this way, Moraes and Moura (2009) point out that educators are guided by the teaching principles developed by scholars of cultural-historical theory and with the aim of elevating the thinking of students, i.e., developing theoretical thinking, we assume that the teacher, when organising their teaching actions that provide opportunities for the appropriation of theoretical knowledge, will also be developing... 102 Bolema, Rio Claro (SP), Ano 22, n° 33, 2009, p. 97 a 116 theoretical knowledge by the students, will also be developing. This perspective is present in the Teaching Guidance Activity when it is seen as a training unit for teachers and students. In this sense, we consider the following main characteristics as the theoretical and methodological basis for organising teaching as an activity: pedagogical intentionality; the existence of a situation that triggers learning; the essence of the concept as the core of the formation of theoretical thinking; mediation as a fundamental condition for the development of the activity; collective work as the context for the production and legitimisation of knowledge. In this way, we believe that the teacher has the important task of organising teaching that has culture as its reference, produced in the development of humanity in such a way as to create meaning for students to appropriate knowledge that allows them to share meanings in their social environment. This task requires objective conditions for its realisation. One of them is the continuous training of teachers, with the core of their studies being their main activity: teaching.

Also according to Moraes and Moura (2009), in the teaching activity, the assessment action has the function of analysing, through the structuring elements of the activity, whether the teaching actions are appropriate to the learning actions, in such a way as to ensure that the learner appropriates the general way of solving the problem situation and transfers it to other situations. In other words, the teacher analyses whether the concept has been appropriated by the students in such a way as to become a symbolic tool in their actions with the surrounding world. Thus, the meaning of assessment in teaching and learning should be to guide and direct the process of appropriating knowledge. This is assessment as an analysis and synthesis of the activities of the subjects, both those who teach and those who learn. In this line of reasoning, assessment should be considered in the relationship between prospection and retrospection, i.e. the subject's previous knowledge is the condition for them to be able to appropriate what is potential for them - what was not available to them at the time - but which, through teaching, they will have the potential to appropriate. This way of assessing the teaching and learning process is an important difference between assessment from a cultural-historical perspective and current pedagogical practices, which only value the knowledge acquired by the student. Culturally, assessment practices have reinforced, through the application of tests and quizzes, what the child already knows, classifying them with the aim of passing and failing. The teaching and learning process is analysed by its product, not by its process. In this case, assessment focussed on the product is partial, since it fails to account for the movement of appropriation of knowledge, to understand what the student is able to achieve with the help of the other - the mediator. Thus, the contributions of this assessment model to educational actions are limited because it does not provide parameters for thinking about the concepts that the student has not yet appropriated. Therefore, if we assume that assessment in cultural-historical theory is a constant process of analysis and synthesis and that its direction is given by the main objective of the teaching activity developed by the teacher - pedagogical intentionality - the teacher, when assessing, will be able to know the process of appropriation of knowledge. In this way, they will be able to intervene in order to guarantee the formation of theoretical knowledge and, consequently, the formation of theoretical thinking.

According to Trevisan, Rocha and Domingues (2017), assessment needs to include the use of different instruments, the selection of which should take place during the teaching planning process, seeking to adapt them to the planned objectives, the established content and the proposed learning activities. In this way, it is essential that "the teacher varies the supports (observation, self-assessment and written records, for example) and the instruments, in order to collect a greater amount of data and thus have information to (re)plan their work and guide student learning". The experiential component associated with "knowing how to assess" and the reflections arising from it are taken as the core of the study: elements of the reflection that took place in (during), on (reviewed outside in its scenario, and documented through the dialogue established between teacher-researcher and supervisor at a time after the experience) and, above all, on the reflection on the assessment action, a retrospective look at the action and the moments of reflection on the assessment action (what happened, the meanings attributed at that moment and others that can be attributed later) are reported and analysed.

In his research, Maciel (2017) mentions that the area of cognitive psychology has made many contributions to making students' cognitive and metacognitive strategies more effective, as well as studying the cause-and-effect relationships that exist in the socio-pedagogical context, including the assessment processes that influence learning and/or the results of eminently summative assessments. This area of knowledge also deals with other psychological variables that are influenced when carrying out an activity, such as motivation, self-esteem, self-efficacy beliefs and achievement goals. Maciel (2017) also reveals that in the context of maths teaching, the consequences of a merely classificatory assessment are extremely serious for the student and for society as a whole. This type of assessment has become very powerful in the hands of teachers of this subject, making it rude, uninteresting, terrifying, among other aspects. The myth that not everyone can learn maths and that summative, purely quantitative assessment has demonstrated this over time, has greatly delayed the development of a didactic approach that would make it possible to give maths its real beauty and importance for the formation of an emancipated citizen, which no one in their right mind can evade. Maths educators have therefore taken into account that 21st century society will demand more from today's students, as they will need to be prepared to solve more challenging problems. They will have to keep up with the evolution of knowledge. Their mathematical communication skills will have to be more efficient.

In the context of teaching and learning maths, from a formative assessment perspective, we need to keep in mind the process of building knowledge individually and as a group. In this environment, students will perceive themselves as part of the class with the same degree of importance as the others. Their relationship with other classmates will be one of co-operation.

Assessment activities will include more than just tests with questions aimed solely at collecting quantitative data on successes and errors. In this way, the assessment of teaching-learning processes requires a diversity of assessment instruments that will reveal to the teacher and the student not only what they are learning, but also how they are learning it. The proposal to diversify assessment tools is not to increase the teacher's workload, but to diversify the way of observing the student and collecting evidence of their mathematical educational development. But every suggestion must be tested so that adaptations can be made according to the educational reality. In any case, in order to regulate learning, it is first and foremost necessary to understand that "the main instrument of all formative assessment is, and will continue to be, the teacher engaged in interaction with the student" (PERRENOUD, 1999, apud MACIEL, 2017). It should be emphasised that the diversification of assessment instruments will only help students learn mathematics if they are detached from the bureaucratic processes of assigning grades. Any value judgement must be the result of the various observations that can be made of the student, and also by the student, with a view to their learning mathematics.

In view of these points about school assessment, the aim of this study was to investigate the meaning of learning assessment for maths teachers and students in the final grades of primary school in the Paraná state public school system.

The aim was to identify the conceptions and purposes of assessment, the most commonly used assessment tools in maths, the difficulties that teachers encounter when assessing and how errors are dealt with in the assessment process.

Considering the complexities that permeate the learning assessment process, we believe that identifying and analysing the opinions and perceptions of teachers and students can provide important food for thought on the subject.

CHAPTER 2

METHODOLOGICAL PROCEDURES

The research methodology adopted was field research, through a qualitative exploratory study.

Twelve public school teachers from the state of Paraná took part in the study, at least two and a maximum of five from each of the following regions of the state: north (Londrina and Maringá), centre-west (Laranjeiras do Sul) and west (Medianeira), who teach maths in the final grades of primary school. As for the students, twelve students from the state public school system in the same regions mentioned above also took part, who are studying in the final grades of primary school and are students of the teachers being researched.

The instruments used for data collection were open-ended questionnaires applied to teachers and students. In order to carry out a qualitative analysis of the information obtained, the Content Analysis Technique proposed by Bardin (1979) was used.

Bardin's Content Analysis (1979) is a set of communication analysis techniques that uses systematic and objective procedures to describe the content of messages, which allowed us to methodologically confront the verification and interpretation of the meanings of the messages (manifest or underlying) attributed to learning assessment by the research participants. In the discussion of the results, an attempt was made to compare the information obtained from the point of view of the teachers and students with the data available in the literature specialising in the assessment of school learning.

The questions posed to the teachers were similar to those posed to the students, as our intention was to see if there was any consensus or disagreement between them with regard to the aspects we consider relevant in learning assessment.

We also carried out a quantitative analysis of the categories that emerged from the pre-analysis and exploration of the material. The quantitative data was complementary, but made it possible to characterise the group of teachers, which made it easier to categorise the responses.

The data was analysed and represented in pie charts in order to better visualise the established percentages of the response patterns acquired through the application of the questionnaire, as well as their quantification according to the answers given to the questions applied to both teachers and students.

CHAPTER 3

DISCUSSION AND RESULTS

The complexity of the elements present in the learning assessment process means that there is no single concept of assessment. In fact, there are different possible ways of approaching the act of assessment. According to Libâneo (1999), assessment is a qualitative analysis of data considered important in the teaching and learning process that helps teachers make decisions about their work.

We agree with Libâneo (1999), but we would also add that assessment should also help the student, because learning assessment also helps them, as it enables them to make decisions about their studies, difficulties and progress. Learning assessment, in this sense, serves both the student and the teacher.

The central focus of our investigation was on the meaning of learning assessment for the teachers and students taking part in the research. The answers given by the twelve teachers were similar in terms of their definition of school learning assessment. This similarity made it possible to group them into a few categories:

- Quantifying student knowledge: 16%

> P1: *"School learning assessment for me is quantifying what the student has learnt, what they haven't and what needs to be revised*

> P6: *"It's a way of seeing how much the student has assimilated of what was worked on in class.*

- Verification of learning: 68%

> P2: *"Assessment is an instrument to verify effective student learning*

> P3: *"And to check the student's interest in the content and their participation during lessons, to see if they have learnt or not*

> P8: *"Assessment allows the teacher to check the student's relationship with knowledge, how they attribute meaning to what they have learnt and how they have been able to apply it in situations that require mathematical reasoning*

> PIO: *"It's about checking what the student has already learnt, it's the moment in the teaching and learning process that allows both the teacher and the student to check whether the proposed objectives have been achieved*

> P4: *"Checking how well my students are doing with the content explained*

- Checking the teaching methodology or the teacher's performance: 16%

> P7: *"Learning assessment provides information so that teachers can reflect on their work in the classroom, and*

> *in order to find more relevant methods that can help students*

P9: "It's a way of checking whether what I'm working on in the classroom is being assimilated by the students, so that if I need to I can change my teaching strategy

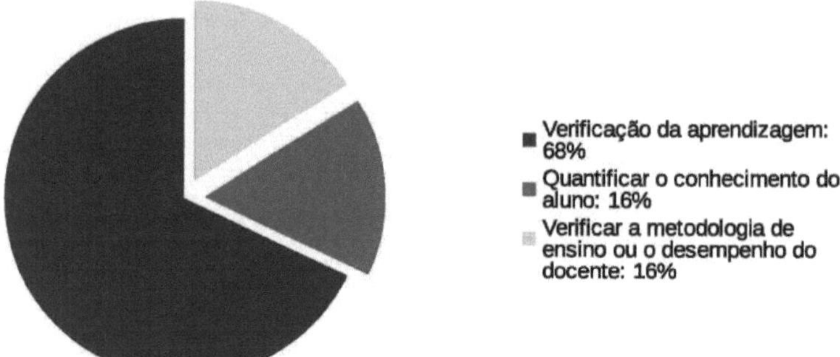

Graph 1 - Graphical representation expressing the school learning assessment response patterns of the twelve teachers interviewed.

In all the teachers' answers, we can see their concern for the student's learning. This shows that although assessment is still considered by the majority of these professionals to be a check on learning, they show between the lines that they care whether the student is learning or not. In this sense, we point to possible changes in the teachers' concept of assessment. Contrary to what many authors say, the teachers interviewed do not see assessment as an end in itself, nor do they see it as punishment, or only for grading students.

According to our analysis and dialogue with these professionals, it is clear that assessment plays an important role in the teaching and learning process.

There was also a concern among the teachers surveyed to reaffirm the need for changes in student assessment and a tendency to broaden their concepts of assessment, going beyond measuring student performance.

In this way, we believe that teachers are trying to fit into the concept of assessment proposed by Luckesi (2002, p. 81):

> *"[...] Assessment should be seen as an instrument for understanding the stage of learning the student is at, with a view to making sufficient and satisfactory decisions so that they can progress in their learning process. If it is important to learn what is taught at school, the role of assessment will be to enable the teacher to understand the stage the student is at, in order to be able to work with him or her so that he or she can move on from the lagging stage he or she is at and make progress in terms of the necessary knowledge [...].*

In this way, the evaluation of learning makes it possible to make decisions and improve the quality of teaching, informing the actions being taken and the need for constant regulation.

Item two of the questionnaire referred to the purposes of evaluation. It was possible to group them into just two categories:

- Verification of student learning: 56%
- Diagnosing possible faults and what needs to be reviewed: 44%

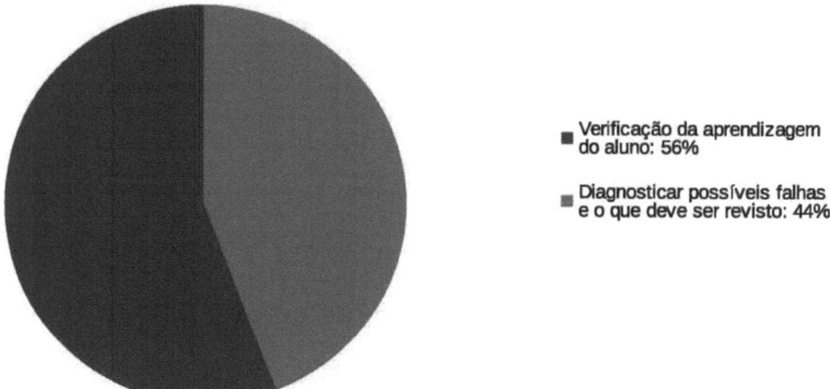

Graph 2 - Graphical representation expressing the responses regarding the purposes of evaluation.

We've selected a few answers that significantly illustrate these categories:

P2: *"Checking what the student has learnt and not learnt"*.

P5: *"Evaluating my own practice, seeing possible flaws and what needs to be revised*

PIO: *"Diagnosing possible teacher and student shortcomings*

P7: *"Check what intervention should be made with the student*

This item showed that many teachers are unclear about the purposes of assessment. In the questionnaire, they were asked to name three purposes of their assessment practices: this was difficult for them to answer, as the majority were unable to name three purposes. Those who did answer generally listed two purposes, which fall into the categories mentioned above.

We can see here that there is a huge gap in terms of the purpose of assessment. We can see from the responses that teachers still don't have enough discernment to identify what the effective purposes of their assessment practices are, and those who do usually refer to the underlying concept of assessment.

Hofftnann (1991) points out that one of the main purposes of assessment is to provide information to improve not only the final product, but also the teaching and learning process. The purposes of assessment practices are deeply intertwined with the concept of assessment that teachers have and use in everyday school life, as well as the aims of the teaching and learning process: what kind of student do you want to train?

The purposes of assessment need to be clear not only to teachers, but also to students. According to Luckesi (2002), the value of assessment lies in the fact that students can find out why they are being assessed and thus check their progress and difficulties. For this reason, it is up to the teacher to first find out what the aims of their assessment practices are so that the student can then also take ownership of this information.

Thus, according to Luckesi (2002), assessment within the learning activity is a necessity for both the teacher and the student. Assessment allows the teacher to acquire the elements of knowledge that will enable him or her to place the action of stimulating and guiding the student in the most correct and effective way possible. It's up to the teacher to challenge them to overcome their difficulties and ensure that they continue to make progress in the construction of knowledge. It is then up to the student to check which aspects they need to improve during their learning process.

In our analysis of the students' answers regarding the assessment of learning, we noticed that many of them confuse the term assessment with testing, making it a synonym for the latter, as well as assigning the teacher the task of analysing and using the results obtained from this assessment process as a priority.

The answers given by the twelve students were very similar in terms of their definition of school learning assessment. This homogeneity made it possible to group them under a single category:

- Checking the student's learning/performance, i.e. the knowledge assimilated or acquired by the student: 99%

Some answers illustrate this category:

> AI: *"These are tests that teachers take to find out how their students are doing. They are assessments of the content that the teacher has explained in class*

> A2: *"They are the assessments made with the student to evaluate what they have learnt from the teacher."*

> A4: *"It's a way of seeing if the students are learning what the teacher is giving them."*

Only one of the students interviewed put assessment from the student's perspective:

> A8: *"[...] it's for us to know if we've understood what we've been told*

According to our observation, the expressions "checking", "knowing" what the student has learnt appeared in practically all the students' definitions, which, like the teachers' answers.

On the question: *What is assessment for?,* the students unanimously (100 per cent) emphasised the aspect of verifying learning.

> A2: *"It serves to show whether the student has learnt a certain piece of content*

> A3: *"For the teacher to see how the student is learning*

> A12: *"To see the individual performance of each student without books, notebooks to help".*

> A10: *"To assess the student in a particular subject and to really see if the student has learnt or not*

According to our readings, the usefulness of assessment is to provide mediated subsidies for correcting the process towards the goal, in this sense educational assessment should mean precisely taking care of the quality of teaching. And this should not only be the teacher's responsibility, but also the student's, as Luckesi (2002, p. 83) states:

> *"The assessment carried out with the students enables the education system to check how it is achieving its objectives, so that in this assessment it has the possibility of self-understanding. If the teacher is attentive to the progress of his or her students, he or she will be able, through learning assessment, to check how efficient his or her work is being and what deviations he or she is having. The student, in turn, will be able to constantly discover what level of learning they are at within their school activity, becoming aware of their limits and their need for progress. In addition, the results shown by means of assessment instruments can help the student in a process of self-motivation, insofar as it makes them aware of the levels of learning achieved".*

In item three of the students' questionnaire, the question asked whether they believed that the school would exist without assessment. The majority (75 per cent) answered no, because they consider the act of assessment to be essential for teachers and schools in order to find out how students are performing and learning. Some selected answers clearly demonstrate this:

> A7: *"No, because without assessment there would be no way of knowing if the student had learnt or not*

> AI2: *"No, because if there was no assessment, hardly anyone would be interested in the subjects in class*

> A2: *"No, because many students don't study even though there are exams, imagine if there weren't. The assessment makes everyone have to revise the subject and study. The assessment makes everyone have to revise and study".*

> A4 *"No, because there would be no way of knowing if the student had learnt. And the student wouldn't study either, and if they didn't study they wouldn't learn. So there would be no point in school, because we go to school to learn and the assessment shows whether we have learnt or not."*

This question was inspired by the text: Evaluation in Everyday School Life, by Maria Tereza Esteban (2003, p. 9), where she asks the following question: *"[...] could the school exist without evaluation? "*

For these students interviewed, we observed that definitely not!

What these answers reveal is that according to Esteban (2003, p. 10):

> *"[...] without the test, what would compel students to study? Without prizes and punishments, with their thousand nicknames and disguises, how could discipline be guaranteed? Without assessment instruments, how to ensure that the minimum content is being learnt, how can we identify who knows and who doesn't, in order to fulfil the school's accreditation function?"*

As Esteban (2002) points out, perhaps assessment is a "necessary evil". In the other answers (25%), even if they answered that the school would exist without assessment, the students were concerned that there would be no learning. This can be seen in the following answers:

> A1: *"I think there could be, but the student wouldn't pay attention and learn because they wouldn't take a test on the subject*

> A2: *"Yes, but the teachers wouldn't know how the student is doing at school, whether or not they've understood the content*

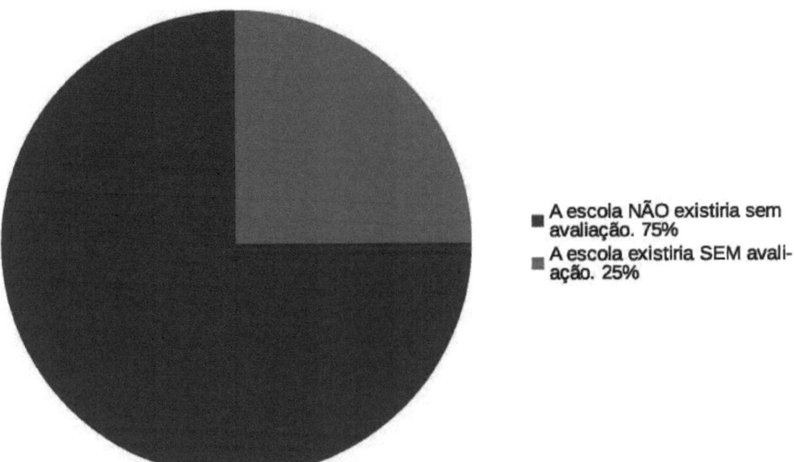

Graph 2 - Graphical representation expressing the students' responses regarding the existence of the school through the presence of assessment.

These students' answers reveal that learning is to a certain extent conditioned to the assessment aspect, i.e. they study to take tests, to get good grades, to finally succeed in their studies and pass the grade. These answers are clear evidence of the role of assessment as a regulator of the teaching and learning process, insofar as the teacher can only guarantee that the student will really study and learn if this is conditional on the assessment aspect. What if there were no tests? What would the students study for? How do you hold the class's attention? How do you get students to pay attention and do the activities if they're not going to be assessed?

In this sense, assessment is a double-edged sword: it is essential for showing the teacher and the student whether the teaching and learning objectives are being achieved, pointing to possible decision-making regarding the results obtained, but it also ends up, in a way, making the students focus their studies and learning on assessment.

We believe this is the great challenge facing schools today: to promote a pedagogical practice that leads students to really learn for life, and not just for exams and schoolwork.

Furthermore, according to Perrenoud (1999), the core function of assessment is to help the student learn and the teacher teach, while also determining how much and at what level the objectives are being achieved. According to Libâneo (1999), this requires the use of appropriate assessment instruments and procedures.

From this perspective, we were also interested in investigating the instruments most used by teachers in their assessment practices.

By analysing the responses, we can see the presence of the following instruments:

Tests (oral, written, individual, in pairs): 100%

Work (individual, in pairs): 100%

These two instruments appeared in all the responses from the teachers surveyed. Some answers clearly show this:

> P2: *"I assess through oral tests, individual tests or tests in pairs. And also through individual or pair work."*

> P3: *"Work in pairs, individual tests or in pairs".*

> P7: *"Work in pairs or individually with the help of the*

teacher, individual tests."

The other instruments mentioned were:

Classroom activities: 42%

Tasks and notebook: 16%

The answers below show these instruments, but they are always associated

with the others: tests and assignments, which are commonly used:

> P4: *"Activities in the classroom and at home, work and*
>
> *activities in pairs, written and oral tests and organisation of*
>
> *activities in the notebook*
>
> Pl: *"Through individual tests, tests in pairs, work with the*
>
> *teacher's intervention, homework assignments, classroom*
>
> *activities, notebook organisation*

These other instruments: classroom activities, assignments and notebooks;

appeared in very few answers, which shows that teachers still stick to the traditional

way of assessing: through tests and assignments. However, it is possible to observe

some nuances of change: oral tests; work in pairs, tests in pairs.

The students were also asked about the assessment instruments and they

confirmed what the teachers had said, which could be grouped into four categories.

Tests (assessment: term used by some students as a synonym for exam): 100%

Work: 100%

Extra and home activities: 100%

Notebook organisation: 33%

The following answers illustrate the students' recognition of the teacher's use of these tools in maths:

> *A2: "In Maths, we are assessed through tests, assignments, notebook organisation, classroom activities, homework on the content being worked on*

> *A7: "I'm assessed by tests, participation in class activities and work based on the content applied*

> *A8: "Tests, individual work, group work, activities".*

> *AI 2: "Through tests, extra tasks and assignments".*

We consider this variety of assessment tools used by teachers to be a step forward, because educators who use different assessment tools to monitor teaching and learning over the course of a term are different from those who just give a test at the end of the term.

However, we must emphasise that this diversity of assessment instruments is also due to the requirement of LDB 9394/96, in article 24, chapter V, as well as Process No. 091/99, deliberation No. 007/99 of the State of Paraná, which obliges teachers to use more than one means of measuring their students.

Luckesi (2002) emphasises the importance of instruments and criteria, because assessment cannot be carried out using data invented by the teacher, who in turn must be clear about the objectives of his assessment practice, the instruments he will use and the criteria that will be analysed for each instrument.

Luckesi (1984) emphasises that the criteria should be used as a demand for quality and not as a form of authoritarianism on the part of the teacher towards the student. Another danger is that the criteria are not formulated in advance but during the assessment itself.

In this context, the decision on the level of learning to be achieved by the student will depend very much on the subjectivity of the teacher when exercising their judgement, because the "mood" of the personality varies not only according to introjected standards, but also according to momentary circumstantial factors.

The task of analysing results initially involves organising all the material, dividing it into parts that must be interrelated and identifying relevant trends and patterns. In learning assessment, the more formalised analysis of results accompanies the application of an instrument and leads to re-planning and guidance for students.

In this way, today's educators must rethink their assessment criteria and the need to build policies and practices that take this diversity into account and that are committed to school success and not failure.

Another question that deserves to be analysed in this article is *what is done when the majority of the class performs poorly in a particular subject or content that has been assessed.*

The teachers unanimously (100%) answered that they retaught the content, but few referred to teaching strategies. They emphasised the issue of doing more exercises and activities on the content that had fallen short.

> PIO: *"I explain again and look for different exercises to apply in class."*
>
> P5: *"After correcting the tests or assignments with the students, the content is explained again, reviewing the content given. Exercises are given so that the student can practise what has been explained, these exercises are corrected and then parallel recovery is carried out so that the students can demonstrate their progress*
>
> P7: *"I go back over the explanations, review the strategies I used and also how the assessment instruments were organised."*
>
> P3: *"Reinforcing the content seen by working with new explanatory methods*

We realise that the majority of teachers, even though they are beginning to point to possible changes, are still very much rooted in the Platonic system of teaching mathematics, as Machado (1987) puts it, which is based on exercises and mechanical activities, without worrying about the aspect of grasping concepts themselves, which in reality are the basis of any mathematical learning. In this sense, teachers end up teaching the way they learned. Mello (2000, p. 98):

> *"[...] no one facilitates the development of what they haven't had the opportunity to improve themselves. No one promotes learning that they haven't mastered.*
>
> *experience as a student what he will have to teach his own students*

From the same perspective, D'Ambrósio (1993, p. 38) points out that: "[...] hardly any maths teacher trained in a traditional programme is prepared to face the challenges of modern curricular proposals". In this respect, we agree with the author and highlight an important question: How can we trust that maths can be learnt differently if teachers have never had such classroom experience as students?

Therefore, we tried to verify the initial training that our teachers had: they all have a degree in Maths and have taken continuing education courses. However, as Ponte (2002, p. 69) points out:

> *"[...] if the competence of teachers were measured by the number of courses attended, the qualification of teachers would be extraordinary. If the quality of schools could be measured by the weight of diplomas and certificates, a revolution would have already taken place in every school. Teachers accumulate "qualifications" without this corresponding to change or responding to the challenges they*

D'Ambrósio (1993) points out that it is very important for maths teachers to have theoretical training so that they know how to deal with new educational trends and thus be able to influence and contribute to the transformation of current assessment practice. Given the training of maths teachers, a lack of theoretical and methodological support, it feels

It's difficult to change assessment practices, because we know that it's hard to change what has been established over centuries.

We won't dwell on this subject of teacher training, but we'll leave you with the following reflection: if the way we teach doesn't change, the way we assess will change even less, because one is interconnected with the other.

The students also answered this question: *what is done when the majority of the class performs poorly in a particular subject or content that has been assessed, and,* like the teachers, they were unanimous *in* answering that the content is retaken. Some answers illustrate this single category:

> A3: *"She goes back over the subject and revises the content and then gives the remedial test to the students who did badly."*

> A6: *"Explain everything again until everyone understands*

A5: "The teacher reviews the content before the exams and if the student does badly, the teacher retakes the content and makes up for it.

A4: "She makes an effort to explain better, asks all the students to take part in the lesson, does more activities on the subject

A9: "She revises the content, does more activities explaining it again

The students' answers further highlight the fact that maths teaching is based on repetition and memorisation. We can see the teacher's effort to ensure that everyone understands the content, but he always does it in the same way: repeating, explaining again, reaffirming his Platonic approach to teaching maths. This finding, based on the responses we received, is extremely worrying, because nowadays this attitude of the teacher explaining, explaining, repeating, repeating, and the students doing and redoing exercises is not enough. In order to "cope well" with the 3rd° Millennium, the 21st century, students need, as Pais (2002) points out, teaching methods that are suited to the growing emphasis on problem-solving, applications and developing deeper reasoning. This is not just done through repetition and memorisation.

In the penultimate question of the content investigated in our research instrument, we sought to find out whether or not there is a treatment of the error made by the student in the assessment tasks and, thus, how the teacher acts in the event of failures and cognitive obstacles that generate conceptual or systematic errors in their students.

The responses from both teachers and students point to the fact that it is not taught in the classroom and that when teachers do teach it, they generally do so inefficiently.

We have grouped them into just one category, as the responses show that the teachers surveyed (100 per cent) work with error from the same perspective: error as a student's cognitive deficiency.

Through the other answers, which were analysed earlier, we can see that the assessment system of which the teachers surveyed are a part, its methods, its assessment instruments, its ways of recovering content, ignore or pay little attention to the attempts that didn't result in an immediate success on the part of each student. Some answers are clear evidence of this:

> PIO: *"The error is pointed out and the student has the opportunity to check what they did wrong, get new explanations and get it right."*

> P9: *"Yes, I do. When I observe their performance in the activities, I take them back and make them realise their mistakes and review them."*

> PI2: *"Errors are corrected on the board and more exercises are given on the subjects that were most missed."*

It is clear from the answers given by the teachers surveyed that they believe the cause of the mistakes made by their students is due to a lack of knowledge, for which they advocate the classic therapy of repeating explanations and exercises. The result, as a rule, is ineffective.

The students' answers also bring out what we observed in the teachers' statements:

> A5: *"He corrects on the board, asks if everyone has understood and then goes from desk to desk making sure everyone has understood the mistake made."*
>
> A3: *"When we make a mistake, the teacher revises the subject again in an easier way so that everyone understands."*
>
> A4: *"During lessons when the subject is being explained, she answers questions, helps us see where we've gone wrong, while in the exam we only get a mark for the exercise we get right*

The importance of knowing the student's mistakes and where they are unable to overcome a particular difficulty lies in the fact that when dealing with mistakes, the teacher has precise data at their disposal for more individualised interventions, as well as recognising their students' differences and difficulties. According to Bertoni (2000,

p. 45), "[...] the most important thing is for the teacher to adopt a reflective attitude towards the student's error, seeking not only to understand the error within a context, but also to understand the subject who errs."

According to Esteban (2003, p. 21):

> *"Error provides new information and raises new questions about the individual and collective learning/development dynamic. Error, often more than success, reveals what the child "knows", placing this knowledge in a procedural perspective, also indicating what they "don't know yet", and therefore what they can "come to know". In this sense, it becomes a stimulus (or a challenge) to the teaching/learning process - a stimulus for those who learn and a stimulus for those who teach. Error reveals the complexity of the knowledge process, woven simultaneously by the past, the present and the future. "*

In this aspect of working with student error, we emphasise that it is of fundamental importance for teachers to review their treatment of student error during the teaching and learning process, especially with regard to assessment practices, trying to find out why the student makes mistakes, the particularities of the cognitive obstacles and the quality of the failures committed, and then acting in the zone of proximal development as pointed out by Vygotski (1994), in other words, intervening where the student, according to Esteban (2002), does not yet know so that they can come to know.

The last question analysed in the questionnaire with the teachers asked them to name the three biggest difficulties they encounter in assessing their students' learning.

It was possible to group the difficulties pointed out into three categories:

- Working with student errors and returning tests and assignments (28 per cent).

> P5: *"Working with students' mistakes, because there are situations in which the student knows how to solve a particular exercise, but makes a mistake due to distraction. Another problem is working with children who have proven difficulties, to what extent should I look at them differently so as not to exclude them, and how do I help them, how do I give them tests and assignments? "*

> PIO: *"The biggest difficulty is working with the feedback from tests and assignments, because there is little time to do a really good job and then failures in learning are realised. "*

- Number of students per class and aspects relating to the students themselves, such as lack of interest (56 per cent).

P7: *"The number of students per class and the very heterogeneous levels of learning and interest of the students in learning."*

P8: *"It's very difficult to assess students on their performance in class, because they're very neglectful, so it's hard to know if they've learnt or not*

- Fair and correct preparation of tests and correct assessment - assessment instruments - (16 per cent).

P2: *"It would be: what level of test to give, how many tests and how many assignments and the most correct way to find out if the student has learnt*

P3: *"Establishing criteria for preparing and marking tests and assignments, as well as how to correctly prepare the instruments we use to assess."*

With regard to the difficulties pointed out by teachers, these indicate how complex the act of assessment is and how many facets it presents.

According to our analysis, a reflective approach to these teachers' pedagogical practice has not yet been noticed. We therefore emphasise the urgent need for teachers to reflect on their teaching practice. According to Hoffinann 1991, reflective practice means turning "inward". It implies knowing that reflecting in order to act means assuming, in practice, the reasoning and spirit of the project; it means valuing action planning and using evaluation as a way of regulating and observing what has not yet happened, but on what we have already defined a value and a way of intervening.

The last question asked of the students that merited our analysis was *whether the form of assessment their maths teacher uses favours their learning*. In their answers, the students unanimously said yes, pointing out that this was because the teacher "explains the subject well" and "uses various ways of assessing".

The answers below illustrate our observations in this regard:

> A4: *"Yes, because she only assesses when she tries to explain until everyone understands*
>
> A7: *"Yes, because she assesses us in various ways."*
>
> A8: *"I think so, because the explanation, the exercises done in class clear up most of the students' doubts. And when we go to the tests, it gets easier."*
>
> A9: *"Yes, because it assesses us in so many ways that it makes our learning more demanding."*
>
> A3: *"Yes, because she doesn't just give tests, but also assignments, different exercises that make you think and*

It is clear from the responses of the students surveyed that the form of assessment used by the teachers is well accepted by the students, as they are already accustomed to this school culture of assessment.

The so-called "assessment culture", which is currently so present in educational processes and which has often conditioned curricular practices, the organisation of school pedagogical work and assessment practices themselves in an attempt to adapt them to the demands of this "culture", has often contributed to the maintenance of certifying school assessment, which is increasingly losing its pedagogical meaning. On the other hand, our concern is to reaffirm the importance that assessment has as a central practice in pedagogical work, but not because it is more valuable than the very purposes of this work, or because these purposes are justified only by the imminence of being assessed. The central importance of assessment lies in the fact that it (should) be able to mediate educational practice, making a significant contribution to achieving its aims and even (re)signifying them.

CHAPTER 4

FINAL CONSIDERATIONS

The discourse of teachers and students reveals how complex and challenging the task of learning assessment is in everyday school life. In the fabric of this discourse we can uncover the place it occupies in the teaching and learning process and the importance it plays in this process.

Aware that there are no ready-made answers, we have tried to respect and carefully analyse the responses we have received. Without pretending to exhaust the studies on assessment, we have tried, as far as possible, to make a cautious analysis based on the literature cited, without pretending to point out solutions for the practice of assessment. We believe it would be inconsistent, to say the least, to list a series of negative aspects that are implications of the evaluation process. However, this was not our intention.

Comparing the answers obtained from the teachers and students surveyed with the specialised literature we had access to only allowed us to make some considerations about school assessment. According to the literature, the learning assessment process includes the continuous collection of quantitative and qualitative data about the extent and nature of student learning. Thus, all the information available on academic performance is useful for both teachers and students because it will help them to judge the value of the results and make decisions about the progress and difficulties that may occur during the teaching and learning process.

This judgement of the value of learning outcomes depends on the teacher's concept of the assessment process, which in turn determines *what and how* they assess. In this way, the possibilities for changes in assessment practices require teachers to realise that different teaching and learning models imply different assessment approaches.

Therefore, reviewing the concept of assessment and the aspects that underlie it certainly means reviewing the concepts of teaching and learning, education and school, based on principles and values that are committed to the institution of the citizen student. When this is put into practice, assessment will be seen as a diagnostic, dialogical and transformative function of school reality.

Change is often a painful process, as traditional assessment procedures and attitudes that are inappropriate for the times in which we live are deeply ingrained in teachers and students. However, it is up to teachers to believe, despite the numerous obstacles, that it is possible to carry out learning assessment in everyday school life with an emphasis on decision-making to ensure the success of the student and consequently of the teacher's pedagogical practice. Because every response to the learning process, whether right or wrong, is a point of arrival, because it shows the knowledge that has already been built up and absorbed, and a new starting point, for a fresh start, enabling new decisions to be made.

However, these decisions must be made in a way that takes into account all the theoretical and methodological procedures that can make a positive and substantial contribution. Because many of the theories already described in books and journals help in a given reality, perhaps one that is close to the one described. However, the teacher, as the articulator of knowledge, must bear in mind that for each reality there is a different perception of the teaching-learning process. The contextualisation of the content is important, but even more important than the content to be studied is the way in which it is approached and the methods used to assess the whole process.

This is why it is necessary to study the different realities, difficulties and facilities at regional and local level, so that territorial and cultural mappings can be studied in order to adequately direct teaching-learning processes so that they can be optimised and put to better use.

In this sense, the study presented here emphasises the importance of on-site studies, studies that reveal regional peculiarities. School assessment is still a much-discussed topic that still needs to be better studied, because the theories that are still used to guide studies come from studies carried out in realities that are different from those experienced today, which is why we reinforce the need for current studies with current realities so that the entire teaching-learning process can be contextualised.

This work was an attempt to systematise a rich set of available data and provide some possible interpretations of the subject. It is hoped that, by disseminating its results, it will have contributed to destabilising crystallised perceptions, as well as to subsidising discussions on the subject, contributing in some way to improving the quality of mathematics teaching in the final grades of primary school in terms of learning assessment practices.

CHAPTER 5

REFERENCES

BARDIN, L. **Content Analysis.** Lisbon: Edições 70, 1979.

BRAZIL. Ministry of Education. National Education Guidelines and Bases Law.
Available at http://www.planalto.gov.br/CCIVIL 03/LEIS/L9394.htm

BERTONI, N. **O erro como estratégia didática.** Campinas: Papirus, 2000.

DAMBRÓSIO, B. S. Maths teacher training for the 21st century: the great challenge.
In: **Pró-Posições.** Campinas-SP: Cortez / UNICAMP, v. 4, n. 1,1993.

ESTEBAN, M. T. A avaliação no cotidiano escolar. *In".*(org.). **Avaliação:** uma
prática em busca de novos sentidos. 4. ed. Rio de Janeiro: DP&A, p. 7-28, 2003.

HOFFMANN, J. **Evaluation Myth & Challenge, a constructivist perspective.**
Porto Alegre: Mediação, 1991.

LIBÂNEO, J.C. Didática. 15.ed. São Paulo: Cortez, 1999.

LUCKESI, Cipriano C. **Evaluation of school learning.** 14. ed. São Paulo: Cortez,
2002.

LUCKESI, C. C. **Avaliação da aprendizagem escolar: estudos e proposições.** 8.
ed. São Paulo:
Cortez, 1998.

LUCKESI, C. C. Avaliação educacional escolar; para além do autoritarismo.
Tecnologia Educacional, Rio de Janeiro: ABT, v.13, p. 6-15, nov./dez., 1984.

MACHADO, N. J. **Matemática e realidade:** análise dos pressupostos filosóficos que fundam o ensino da matemática. São Paulo: Cortez, 1987.

MELLO, G.N. Initial teacher training for basic education: a radical (re)vision. **São Paulo in Perspective Magazine.** São Paulo: SEADE, vol. 14, n.l, p. 98-110, jan./mar. 2000.

PAIS, L.C. **Didática da Matemática:** uma análise **da** influência francesa. 2. ed. Belo Horizonte: Autêntica, 2002.

PERRENOUD, P. **Avaliação:** da excelência a regulação das aprendizagens. Porto Alegre: Artes Médicas Sul, 1999.

PONTE, J. P. Investigating our own practice. In: **Reflecting and investigating professional practice.** Lisbon: APM, p. 5-28, 2002.

VYGOTSKY, L. S. **A formação social da mente.** 2 ed. São Paulo: Martins Fontes, 1994.

SANTOS, L. The **assessment of learning in Maths: A look at its journey.** Available at: http://www.esev.ipv.pt/matlciclo/2007%202008/ temas %20matematicos/apa.pdf. Accessed on 08/02/2018.

MORAES, S.P.G.; MOURA, M. O. **Avaliação do Processo de Ensino e Aprendizagem em Matemática: contribuições da teoria histórico-cultural.** Bolema-mathematics education bulletin-boletim de educacao matematica, v.22, n.33, p.97-116, 2009.

TREVISAN, A. L.; ROCHA, Z. F. D. C.; DOMINGUES, N. D. **Analysing a first experience with phased testing: reflections in, on and about reflection in assessment practice.Educação** matemática em Revista ISSN 2317-904, 2017.

MACIEL, D. M. **Formative assessment and metacognitive assessment tools in maths education: an effective aid to teaching and learning.** Educação matemática em Revista ISSN 2317-904X, 2017.

Printed by Books on Demand GmbH, Norderstedt / Germany